SPOTLIGHT ON NATURE

# WHALE

NELL MUSOLF

CREATIVE EDUCATION · CREATIVE PAPERBACKS

Published by Creative Education and Creative Paperbacks
P.O. Box 227, Mankato, Minnesota 56002
Creative Education and Creative Paperbacks
are imprints of The Creative Company
www.thecreativecompany.us

Design by Blue Design, Inc.
Art direction by Tom Morgan
Edited by Grace Beltowski

Photographs by flickr/Biodiversity Heritage Library, 8, 10, 14, 16, 20, 22; Getty Images/by wildestanimal, cover, 1, Francois Gohier/VW Pics, 10, Gerard Soury, 12, Reinhard Dirscherl/ullstein bild, 6, Sean Gallup, 21, Stuart Westmorland, 24, Tfilm, 11; Unsplash/Chinh Le Duc, 18, Mendar Bouchali, 28, Richard Sagredo, 21, Thomas Kelley, 14, Ulrike R. Donohue, 3, 29; Wikimedia Commons/Ed Lyman/NOAA/National Marine Sanctuaries, 29, Gregory "Slobirdr" Smith, 9, Jérémie Silvestro, 27, National Marine Sanctuaries, 15, 17, 23, 29, NOAA, 16, Sylke Rohrlach, 4–5, Will Falcon aka Vitaly Sokol, 29



Library of Congress Cataloging-in-Publication Data
Names: Musolf, Nell, author.
Title: Whale / Nell Musolf.
Description: Mankato, Minnesota : Creative Education / Creative Paperbacks, [2026] | Series: Spotlight on nature | Includes bibliographical references and index. | Audience: Ages 10-13 | Audience: Grades 4-6 | Summary: "Submerge yourself in the whale's world with this nature title for middle-grade wildlife lovers. Informational text pairs with a narrative about a single whale family to spotlight the wild marine animal's life cycle, supported by infographics and a timeline of developmental milestones"— Provided by publisher.
Identifiers: LCCN 2024048413 (print) | LCCN 2024048414 (ebook) | ISBN 9798889896296 (library binding) | ISBN 9781682777954 (paperback) | ISBN 9798889897095 (ebook)
Subjects: LCSH: Whales—Juvenile literature.
Classification: LCC QL737.C4 M87 2026 (print) | LCC QL737.C4 (ebook) | DDC 599.5—dc23/eng/20241223
LC record available at https://lccn.loc.gov/2024048413
LC ebook record available at https://lccn.loc.gov/2024048414

Printed in India

# CONTENTS

MEET THE FAMILY

# HUMPBACK WHALES

## of the Azores Islands

It's a hot August day in the Azores Islands. This paradise made up of nine islands is located 1,000 miles (1,609 kilometers) west of Portugal in the North Atlantic Ocean. The air above the islands is warm, but beneath the ocean surface the water is a mild 70 degrees Fahrenheit (21.1 degrees Celsius). The world's oceans are a vast space where more than 260,000 different plant and animal species live. The humpback whale is one of them.

A female humpback whale has been waiting to give birth for more than a year. A group of other female whales gathers around her. They are acting as **midwives**. They will help the mother give birth to a **calf**. The whales have chosen a spot in shallow water, close to the shore. This location will help the newborn get to the surface quickly after birth. Until then, the whales wait.

CLOSE-UP

## Rubbery skin

The skin on a baby whale is smooth and feels like wet rubber. A whale's skin remains smooth as it matures. It can get bumpy from parasites, jostling with other marine creatures during feeding, or hitting a boat in the water.

CHAPTER ONE

# LIFE BEGINS

Whales are found in all the world's oceans. They live in the warm waters of the Indian Ocean to the freezing depths of the Arctic. There are over 90 different species of whales. They vary in size and color, yet they all share key characteristics. They are **warm-blooded**. They have hair. Female whales give birth and nurse their young. All whales spend their lives mostly underwater, coming up to the surface for air.

Unlike fish, whales can't breathe underwater. Fish have **gills**. Whales breathe air using their lungs and a blowhole located on top of their heads. Whales expose their heads to get air when they are near the water's surface. After taking a breath, the blowhole seals tightly to prevent water from entering. Deep divers like sperm whales can last underwater for more than an hour without taking a breath.

There are two types of whales: baleen and toothed. Baleen whales have plates in their mouths. When baleen whales collect seawater in their

HUMPBACK WHALE MILESTONES

DAY 1

- Born
- Weight: 2,000 pounds (907 kilograms)
- Length: 13 feet (3.9 meters)
- Can swim a few minutes after birth
- Nurses on mother's milk

mouths, the plates separate the water from prey such as **krill** and small fish. Toothed whales use their teeth to catch larger sea creatures, including fish and squid.

Baleen whales are known as great whales because of their size. The largest baleen is the blue whale. Blue whales are the largest mammals on the planet and can grow to 108 feet (33 m) long. They weigh between 50 and 150 tons (45–136 metric tons). The largest known blue whale weighed 209 tons (190 t). The most common baleen is the minke whale. There are almost one million minke whales living in the North Atlantic, Northwest Pacific, and Antarctic oceans.

Toothed whales include sperm, beluga, and beaked whales. They use **echolocation** to help with navigation and hunting for food. They blow air through their nasal passages. This creates a high-pitched sound that

CLOSE-UP

## Tubercles

Humpback whales have bumps on their heads, mouths, and flippers called tubercles. Tubercles are large hair follicles, similar to whiskers. Each follicle contains nerve endings and a single hair. Researchers believe tubercles sense the vibrations of other animals in the water.

FEATURED FAMILY

## Welcome to the World

The female humpback whale is ready to give birth. The birthing process begins, and the other whales form a circle around the mother. The calf's tail emerges first, quickly followed by the rest of its body. When the calf is completely born, it is one quarter the length of its mother. A newborn whale can't swim for a few minutes after birth. The other whales help the mother support her calf. Together, they begin to nudge the calf upward. They reach the surface of the water. The calf takes its first breath.

bounces out into the ocean environment before coming back to the whale. The sound gives the whale information about the surrounding water, such as approaching predators.

Baleen whales hear low-pitched sounds. These sounds travel far distances and cover large areas of water. Whales communicate with each other using these low-pitched sounds. The noises are known as whale songs since it sounds as if the whales are singing to each other.

**6 MONTHS**

- Watched closely by mother
- Continues nursing on mother's milk
- Drinks between 2 and 10 percent of weight in milk daily

CLOSE-UP

## Color blind

Whales can tell light from dark, but they can't tell the difference between red and green. Their eyes are **monochromatic** and contain one type of cone. The cone corresponds to a color between red and green. The exact color a whale sees depends upon its species.

FEATURED FAMILY

## First Meal

The newborn humpback whale is ready for his first meal. Like almost all mammals, whales live on their mother's milk until they can hunt for themselves. Unlike other mammals, humpback calves do not latch onto their mothers. The mother carefully positions the calf over herself. She then squirts her milk into the calf's mouth. The milk is thick and full of fat, which keeps it from separating in the water. The milk's nutrients will give the calf everything he needs to grow strong and healthy.

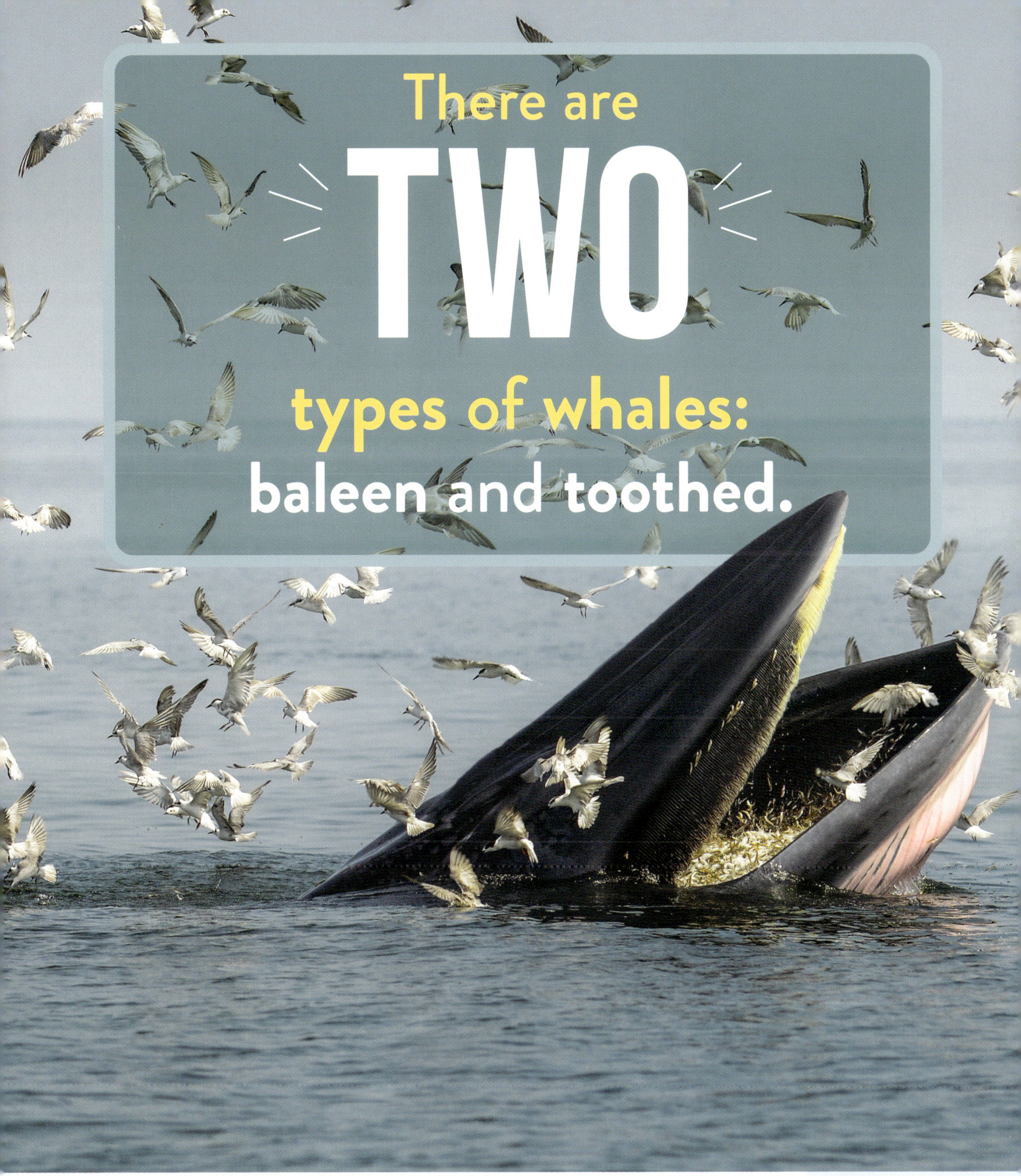

# There are TWO types of whales: baleen and toothed.

## 1 YEAR

- Socializes with other whales from same species
- Migrates to warmer waters for first time
- Length: 26 feet (7.9 m)
- Weight: 8,000 pounds (3,629 kg)

CLOSE-UP

## Diatoms

Blue whales aren't actually blue—they are gray. When a blue whale comes to the surface of the ocean for air, **diatoms** on its skin reflect the color of the water, giving it a blue appearance.

CHAPTER TWO

# EARLY ADVENTURES

Newborn whales often don't sleep for the first month of their lives. Mother whales also stay awake to keep an eye on their offspring. A whale calf begins nursing shortly after birth. The calf might nurse up to age three, but most are **weaned** by their second year. Whales stick closely to their mothers while learning how to navigate the ocean. A calf will swim next to its mother and learn how to **breach** the water by copying her moves. The more the calf practices, the better it gets at moving through the water.

Whales communicate with their calves before they are born. Each whale has a special whistle it uses with its offspring. Once the calf is born, the whistle also helps the mother whale identify her calf in the pod. A pod is a group of whales that are usually related and travel together. Calves communicate with their mothers by softly calling out to them. The soft tone of the calf protects it from being heard by predators. As calves mature, they learn how to make louder vocalizations, or whale songs. They

**2 YEARS**

- Drinks less of mother's milk
- Starts to eat krill and small fish
- Hunts with other whales

CLOSE-UP

## Flippers

The front limbs of a whale are its pectoral flippers. These flippers help the whale steer its body to navigate the water. The flippers also help the whale come to a stop.

FEATURED FAMILY

## Eat Up!

It's mealtime for the humpback whale and her calf. Venturing farther into the ocean, they swim side by side. Mom spots a patch of krill. With her mouth open, she rotates counterclockwise, just below the surface. Then she lunges through the water. Her mouth fills with dozens of tiny krill. The calf closely follows what his mother is doing. He opens his mouth, rotates his body, and then lunges. He makes a successful catch!

communicate with other whales in the pod. They also use vocalizations to warn of possible danger.

The first year of a calf's life can be difficult. Predators such as sharks prey on weaker calves that can't swim as fast as adult whales. After their first year, whales become more skilled at avoiding predators. They are bigger and stronger. They can move faster and they know their surroundings better.

**3 YEARS**

- Fully weaned from mother
- Ventures farther away from mother
- Seeks a new pod with other whales

CLOSE-UP

## Many stomachs

Most whales have three to four stomachs. Baird's whales, the largest member of the beaked whale family, have 13 stomachs. Since whales typically eat prey whole, these multiple stomachs help whales break down and digest food. The process takes 15 to 18 hours.

FEATURED FAMILY

## Give It a Try

The humpback whale is two months old. He can swim and is learning to feed himself. He's getting more independent but still has a strong bond with his mother. Every day they swim together. The calf swims in the wake his mother leaves behind. Following his mother's wake makes it easier for the calf to swim and helps him develop his muscles. When he is strong enough, he will be able to swim just as fast as his mother—up to 16 miles (25.7 km) per hour.

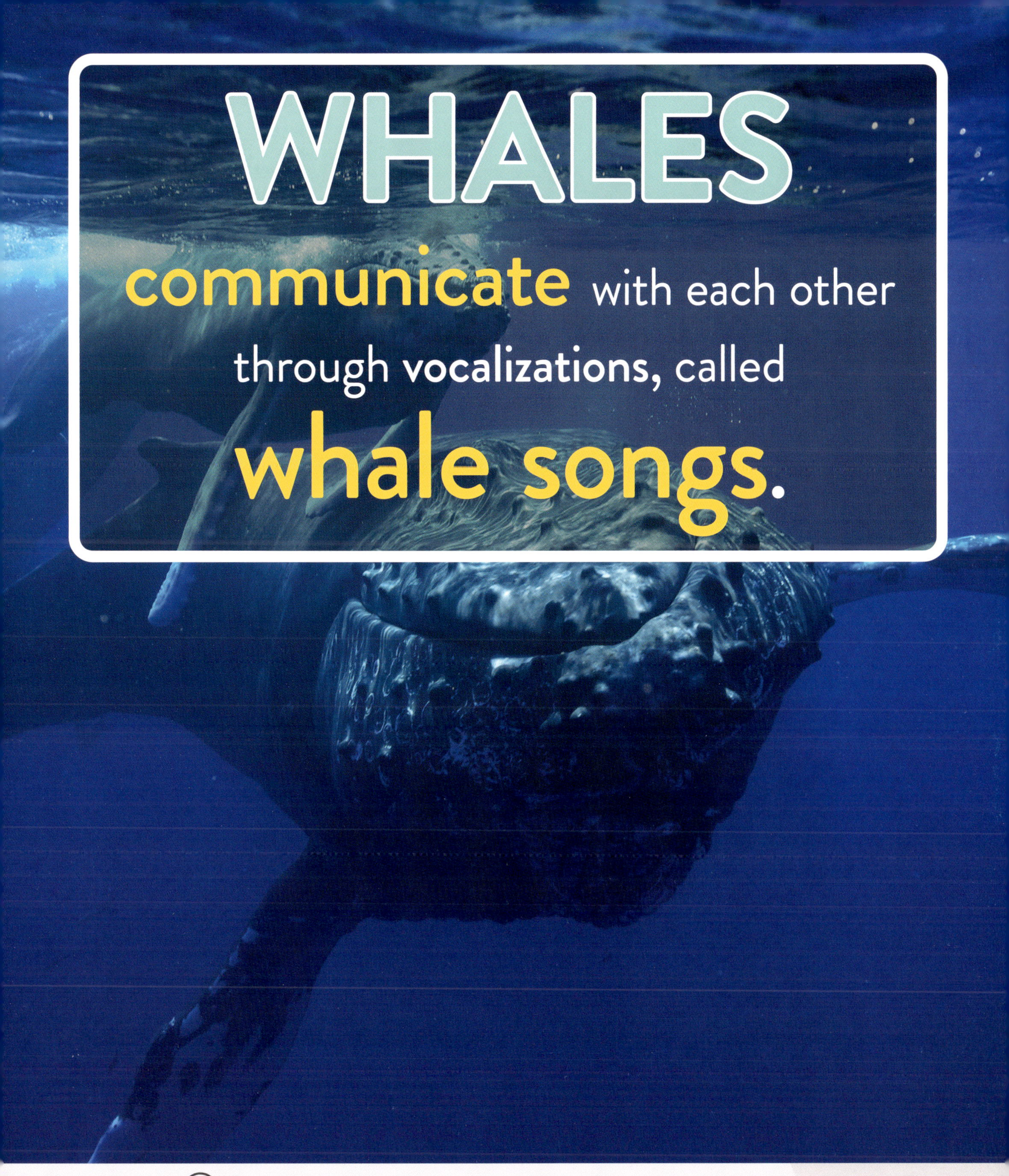

# WHALES

**communicate** with each other through **vocalizations,** called **whale songs.**

## 6 YEARS

- Reaches sexual maturity
- Mates for the first time
- Weight: 32 tons (29 t)
- Length: 45 feet (13.7 m)

CLOSE-UP

## Special songs

Each mating season, male humpback whales create a song that can last as long as 20 minutes. The song is repeated for hours and all the males in the pod sing the same one. A new song is created the next season.

CHAPTER THREE

# LIFE LESSONS

Whales often help each other find food. They sometimes hunt together and form a net made up of air bubbles from their blowholes. Prey is caught in the net and then shared among the group. Whales are social animals. As they get older, they may seek the company of other whales of the same species, age, and gender. Some males leave their mothers and join pods made up of other males. They begin looking for females to mate with them. Female whales also leave their pods and meet other whales, but they usually return to their mothers after the mating season ends.

Mating occurs during **migration** to warmer waters in the winter. Whales migrate when ocean water becomes cold and food sources are scarce. They seek warmer waters closer to the equator where there is more food. During this time, males attract females in many ways. They blow bubbles, sing, breach, and sometimes fight with other males to get attention from females. When whales become a couple, they swim together and mimic

**10 YEARS**

- Fully grown
- Weight: 38 tons (34.5 t)
- Length: 50 feet (15.2 m)
- Females give birth every two to three years

each other by diving and rolling in the water. Female whales are pregnant for about a year. They give birth every two to three years.

Whales have a long lifespan that varies between species. Some, such as beluga whales, live up to 60 years. Others, including the humpback, live up to 90 years. The Bowhead whale lives the longest of all whale species with an average lifespan between 100 and 200 years.

A whale's habitat can impact how long it lives. Whales living in **captivity** have much shorter lifespans than whales living in the wild. Living in the smaller spaces of captivity can cause health problems such as skin conditions that shorten the whale's lifespan.

One reason it is important to keep whales in their natural habitat is because they play an important role in the health of oceans. They bring

FEATURED FAMILY

## This Is How It's Done

Hey, you there! The young whale is swimming with his pod when he spots a female whale he'd like to get to know. He begins blowing bubbles through his blowhole. Then he breaches the water, leaping into the air to show off how high he can go. After he's back in the water, he begins singing. All his showing off works. The female whale notices him and swims over. Together they start swimming side by side. They are now a couple.

CLOSE-UP

## Blubber

Whales have a thick layer of blubber under their skin that keeps them warm in frigid ocean water. Blubber is also where energy is stored for future use when food is unavailable. Blubber can make up as much as 50 percent of a whale's entire body weight.

### (40) YEARS

- Migrates to warmer waters each winter
- Skin has bumps from hitting other animals and objects in water

nutrients to other parts of the ocean when they migrate. They are also able to store large amounts of carbon in their bodies because of their massive size. They get this carbon by eating carbon-rich prey. Storing carbon helps remove it from the atmosphere. Less carbon in the atmosphere helps maintain global temperatures.

CLOSE-UP

## Humpbacked

Humpback whales got their name from the humps in front of their dorsal fins. These fins are shorter than those of other whales. They make the humps more visible when the whales arch their backs to dive in the water.

FEATURED FAMILY

## Practice Makes Perfect

The whale joins other males in his pod, and together they swim out into the ocean. There they spot a school of silvery capelins, small fish found in northern oceans. Forming a circle, a few whales start blowing bubbles into the water. Quickly, a curtain of bubbles surrounds the capelins. One whale signals to the others. Then, they all move forward quickly with their mouths open. The school of capelins becomes a tasty meal.

# WHALES

## migrate to warmer waters in the winter.

**85 YEARS**

- End of life

CHAPTER FOUR

# HELPING WHALES SURVIVE

For thousands of years, whales have been hunted and killed by humans. This practice is known as whaling. Whale oil was used to make candles and burned in lamps before electricity. Baleen, the plate in some whale species used to strain prey from seawater, was also valuable to humans. Baleen was used in clothing such as collars to make them stiff. Another whale byproduct was ambergris. Ambergris came from the intestines of sperm whales and was used in some perfumes.

In 1986, commercial whaling was banned by the International Whaling Commission. However, some countries still allow whales to be hunted, including Norway, Iceland, and Japan. Illegal whaling is also an issue around the world.

Some whale species are endangered, such as the North Atlantic right whale. Less than 400 whales of this species remain, largely due to whaling. Another threat all whale species face is climate change. Climate change raises the temperature of sea water. As ocean temperatures rise, many food sources for whales, such as krill, could become less available or disappear. If climate change happens rapidly, whales and other marine animals may not have time to adapt.

Whales face other threats. They can become entangled in fishing equipment and marine debris, such as garbage thrown from ships and lost fishing gear. When entangled, whales can starve or drown. Humpback whales often collide with boats, causing injury to their skin. Collisions between a whale and a sea vessel are dangerous for everyone involved.

There are many organizations that help all ocean life, including whales. The Marine Conservation Society cleans up trash on beaches in Great Britain. Sea Shepherd monitors the oceans with patrol boats looking for illegal activities, including illegal whaling. Oceana works toward creating science-based policies to stop overfishing, protect ocean habitats, and decrease plastic pollution. Donating to these organizations or volunteering with a locally based environmental group are great ways to help protect whales and all other sea creatures.

Whales are magnificent creatures that deserve to live in a safe, clean ocean. Taking action to keep oceans clean is the first step in ensuring whales are around for a long time to come.

FAMILY ALBUM

# SNAPSHOTS

**Blue whales** gain 200 pounds (90.7 kg) every day for the first year of their lives.

The heart of a blue whale is the size of a small car. It pumps 10 tons (9.1 t) of blood every day and has an aorta that is big enough for a human to crawl through.

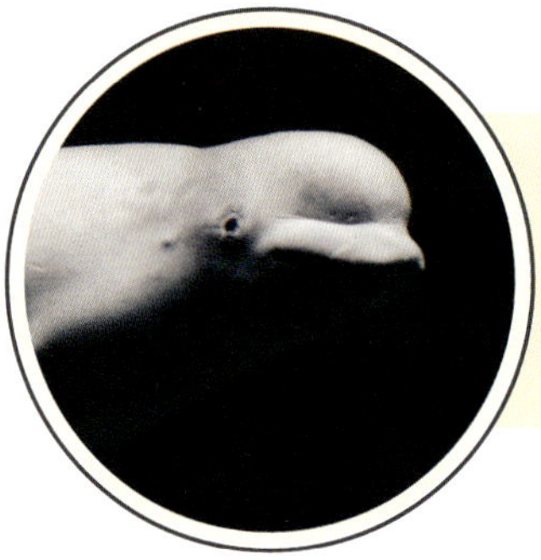

**Beluga whales** have been nicknamed the "canary of the sea" because of the many different noises they can make.

Beluga whales can swim backwards. They also have neck bones that are not fused together, which allows them to move their heads in more directions than other whale species.

A **Cuvier's beaked whale** went for 137 minutes between breaths. This is the known record for the longest amount of time between breaths for a whale.

The **sperm whale** has the largest brain of any mammal. Its brain can weigh up to 9.2 pounds (4.2 kg).

The blowholes of different whale species have different shapes. **Humpback whales** have two separate blowholes.

**Gray whales** have a single V-shaped blowhole.

Humpback whales eat up to 5,500 pounds (2,495 kg) of krill and fish every day.

**Minke whales** are the smallest baleen whales. They grow to be about 35 feet (10.7 m) in length and weigh up to 10 tons (9.1 t).

Minke whales in the Antarctic have pectoral flippers that are solid gray with white edges.

# WORDS to Know

| | |
|---|---|
| **breach** | rise and break through the surface of the water |
| **calf** | a baby whale |
| **captivity** | a state of imprisonment or confinement |
| **diatom** | a microscopic alga on the surface of a whale's skin that reflects light |
| **echolocation** | the process of locating objects by reflected sound waves |
| **gill** | the respiratory organ used by fish to draw oxygen from the water |
| **krill** | a small shrimp-like crustacean eaten by whales |
| **midwife** | someone, usually female, who helps with childbirth |
| **migration** | the seasonal movement of animals from one place to another, usually for food and breeding |
| **monochromatic** | containing or using only one color |
| **parasite** | an animal or plant that survives on or within a host organism |
| **wake** | a disturbance of water causing waves behind it |
| **warm-blooded** | able to naturally maintain one's own body temperature, typically above that of the surroundings |
| **wean** | accustom an infant to food other than its mother's milk |

# LEARN MORE

## Books

Kirkman, Marissa. *Humpback Whales*. Mendota Heights, Minn.: Apex Editions, 2025.

O'Brien, Cynthia. *Whale, Dolphin, and Other Pods*. St. Catharines, Ont.; New York: Crabtree Publishing, 2023.

Skerry, Brian, and Stephanie Warren Drimmer. *The Ultimate Book of Whales*. Washington, D.C.: National Geographic Kids, 2025.

## Websites

"Blue Whale." Ducksters.
https://www.ducksters.com/animals/bluewhale.php

"Humpback Whale Facts!" National Geographic Kids.
https://www.natgeokids.com/uk/discover/animals/sea-life/humpback-song/

"Whale." Britannica Kids.
https://kids.britannica.com/students/article/whale/277718

## Documentaries

Armstrong, Brian, and Andy Mitchell, dirs. *Secrets of the Whales*. New York: National Geographic, 2021.

Dougherty, Michelle, and Daniel Hinerfeld, dirs. *Sonic Sea*. Los Angeles: Imaginary Forces, 2016.

MacGillivray, Greg, dir. *Humpback Whales*. Laguna Beach, Calif.: MacGillivray Freeman Films, 2015.

Note: Every effort has been made to ensure that any websites listed above were active at the time of publication. However, because of the nature of the Internet, it is impossible to guarantee that these sites will remain active indefinitely or that their contents will not be altered.

# Visit

### LIME KILN POINT STATE PARK

*Walk along the shore for a chance to see whales in the wild. May to September are the best viewing times.*

1567 Westside Road
Friday Harbor, WA 98250

### MYSTIC AQUARIUM

*Visit beluga whales in person or schedule a private Zoom call with a beluga and its care team!*

55 Coogan Boulevard
Mystic, CT 06355

### MONTEREY BAY AQUARIUM

*From April to December, humpbacks can sometimes be spotted from the aquarium's ocean-view decks.*

886 Cannery Row
Monterey, CA 93940

### SHEDD AQUARIUM

*This aquarium is home to eight beluga whales that are always busy playing and making noises.*

1200 South Lake Shore Drive
Chicago, IL 60605

# INDEX